S-h-h-h-h! Listen closely. What sounds do you hear?

You might hear soft sounds, like the rustling of leaves . . .

or wind blowing through trees.

You might hear loud sounds, like a fire truck's siren . . .

or the crashing of waves.

We hear with our ears.
Our eardrums vibrate as
they pick up sounds.

When something vibrates, it moves back and forth very quickly. Strings on a harp vibrate.

Some sounds are pleasant, like music.

Some sounds are too noisy!

You can make sounds, too.
You can make soft sounds,
like whispering.

You can make louder sounds, like talking and laughing.

Animals also make a lot of sounds. What sounds do these animals make?

10

11

Sound can travel from place to place. It can travel through air.

It can travel through solids.

It can even travel through water.

Some sounds give us warnings.

Some sounds give us information.

And some sounds are just fun!